Northern Lights

Northern Lights

by D. M. Souza

NATURE IN ACTION

Carolrhoda Books, Inc./Minneapolis

The southern lights as seen from Spacelab III ▶

METRIC CONVERSION CHART		
To find measurements that are almost equal		
WHEN YOU KNOW:	MULTIPLY BY:	TO FIND:
miles	1.61	kilometers

Text copyright © 1994 by D. M. Souza
Series editor: Marybeth Lorbiecki
Illustration on p. 9 (bottom), diagrams, and maps by
Laura Westlund, © 1994 Carolrhoda Books, Inc.

Library of Congress Cataloging-in-Publication Data
Souza, D. M. (Dorothy M.)
    Northern lights / by D. M. Souza.
        p.   c.m — (Nature in action)
    Includes index.
    Summary: Discusses the origins, characteristics, and
lore of the Northern and Southern Lights known as
auroras.
    ISBN 0-87614-799-6
    1. Auroras—Juvenile literature. [1. Auroras.] I. Title.
II. Series: Nature in action (Minneapolis, Minn.)
QC971.4.S68   1994
538'.768—dc20
                                        93-3027
                                            CIP
                                             AC

Manufactured in the United States of America

1   2   3   4   5   6   -   P/JR   -   99   98   97   96   95   94

Special thanks to Dr. William J. Stringer, from the Geophysical Institute of the University of Alaska, Fairbanks, for his assistance with this manuscript. Thanks also to the staff of the Geophysical Institute; James Connor, a graduate student at the institute; Dr. David Evans at NOAA; Professor Steve A. Jacobson at the Alaska Native Language Center; Darlene Orr at Carrie McLain Museum of Nome; and the University of Alaska Press.

Photographs courtesy of: front cover, pp. 2, 23 (top), 25 (bottom), 29, 30 (left), 37, 43, 44, 48, © James Connor, Geophysical Institute; back cover, pp. 8, 45 (left), © Eskimo Museum, Bernard Frasen, Roman Catholic Diocese of Hudson Bay; pp. 1, 25 (top), © John & Ann Mahan; p. 4, © R. Overmyer/NASA; pp. 5, 31, © Grant Klotz, New England Stock Photo; p. 6, © Dr. Syun Akasofu, Geophysical Institute; p. 7, Glenn Allred, Space Dynamics Lab, Utah State University; pp. 9 (left), 45 (right), *Aurora: Their Character and Spectra,* by J. Rand Capron, 1879; pp. 10–11, © Brian Donovan; pp. 12–13, 22, Finnish Tourist Board; pp. 14, 16, 20, 33, 34, 40, 41, NASA; pp. 17, 28 (bottom), Kathy Raskob, IPS; pp. 23 (bottom), 24 (right), 39, © F. T. Berkey, Utah State University; pp. 24 (left), 35, 38, © Dr. L. A. Frank; pp. 26–27, Robert Eather; p. 27, © John Sohlden, Visuals Unlimited; p. 28 (top), Amy Cox, IPS; pp. 30–31, © George Cresswell, Geophysical Institute; p. 42 (both), NOAA/SEL; p. 46, John S. Foster.

To Natalie

Contents

Opening   6

What Causes an Aurora?   12

   Solar-Wind Particles   14

   Magnetic Fields   17

   Gases in the Atmosphere   21

What Do Auroras Look Like?   22

   Shapes and Movements   24

   Colors   28

When Are Auroras Most Brilliant?   32

   Sunspots   32

   Solar Flares   34

When and Where Can the Northern Lights Be Seen?   35

   During a Quiet Sun   38

   During an Active Sun   39

   Danger Signals   40

Studying Auroras   41

Tips for Spotting Northern Lights   43

Fascinating Facts   45

Glossary   47

Index   48

A faint, greenish glow suddenly appears in the night sky. It brightens into huge ribbons of light that stretch from east to west, as far as the eyes can see.

6

Red and green rays flash within the ribbons. They sway and flicker like flames in a campfire. Then toward morning, these colors simply fade in the light of dawn.

7

People living in parts of Alaska, Canada, Greenland, Iceland, Sweden, Denmark, Norway, and Finland see these lights frequently. For hundreds of years, they wondered what was happening in the sky.

Swedish people thought the lights were merry dancers. They spoke of the lights as "polkas," a type of folk dance. People who lived in the area of Finland said foxes with sparkling fur were running over the mountains of Lapland (the land north of them within the Arctic Circle). They called the lights "fox fires." Athabaskan peoples of Alaska thought the spirits of the dead who watched over them, or "sky dwellers," were sending them messages.

An Inuit sculpture showing two old men watching a spectacular display of the northern lights

In 1621, a French scientist, Pierre Gassendi, saw the lights in the north and named them after the Roman goddess of dawn, Aurora. He added the word "borealis" for the Roman god of the north wind, Boreas. The lights became known to scientists as the *aurora borealis*.

During a polar voyage in 1861, Dr. Isaac Hayes illustrated the northern lights *(top)*. Aurora, the goddess of dawn, pours out the morning dew in this copy of a picture from a Greek vase *(right)*.

The southern lights over
Auckland, New Zealand

More than a hundred years later, Captain John Cook saw the aurora borealis while he was sailing from Hawaii to Alaska. When he explored Antarctica in 1773, he spotted similar lights in the sky. He named them *aurora australis,* or the southern dawn.

We now call the auroras the northern and southern lights. This book looks most closely at the northern lights, for these are the lights people in parts of the northern hemisphere may see.

What Causes an Aurora?

Stretching hundreds of miles above our planet is a huge ocean of air. It is known as the atmosphere. It contains many gases, but the main ones are nitrogen and oxygen.

The northern and southern lights are caused by high-speed particles from the Sun striking these gases. The particles make the gases glow.

The action is similar to what happens when you turn on a television. At one end of the picture tube is a source of electricity, which is made up of tiny invisible particles called electrons. As the electrons enter one end of the tube, magnets pull them to the other end. There they strike chemicals on the back of the television screen, making them shine. Images then appear on the screen.

During an aurora, the skies become like giant TV screens. The Earth, which is an enormous magnet, pulls particles from the Sun toward the skies above its magnetic poles, which are near the North and South Poles. There, the particles strike gases, and the gases glow like the chemicals on the back of a TV screen. Shimmering lights dance across the dark night sky.

For every aurora, then, three things are needed: particles from the Sun, a powerful magnet, and gases.

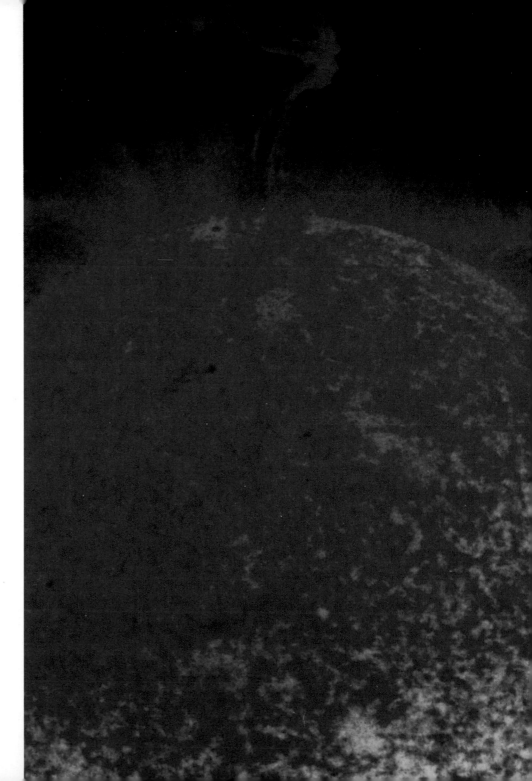

Solar-Wind Particles

Our Sun is a huge bundle of fiery gases. It is so large that more than a million Earths could fit inside it. Temperatures at the center are 118,000 times hotter than boiling water. This extreme heat turns gases into a thin, hot soup of broken-up atoms.

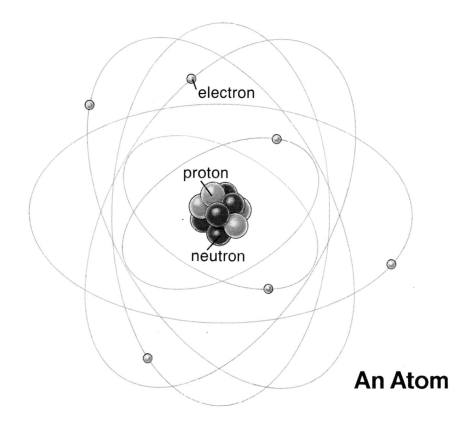

An Atom

Everything in the universe, whether it is a solid, a liquid, or a gas, is made up of parts called atoms. Atoms are so small that over two billion fit inside the period at the end of this sentence.

As small as they are, atoms contain even tinier particles. Some, called protons and neutrons, bind together to form the center of the atom. Others, known as electrons, swirl around the center.

(Remember, electrons also race through wires giving us electricity.)

Under the heat of the Sun, the atoms in the Sun's gases cannot hold together. Some protons, neutrons, and electrons on the Sun's surface escape and stream through space in all directions. They form an outward-moving flow of particles known as solar wind.

Solar wind travels at 200 to 500 miles a second. The spacecraft *Mariner* first tracked the particles and measured their speed in 1962. They were moving 28 times faster than the spacecraft itself.

The solar-wind particles lash against planets as close to the Sun as Mercury and Venus, and as far away as Neptune. Some head toward Earth—93 million miles away from the Sun. About two days later, they slam into a gigantic shield thousands of miles above us in space. The shield is the edge of Earth's magnetic field.

Solar-wind particles are drawn by Earth's magnetic pull toward its magnetic poles. This satellite photograph shows the resulting aurora above Earth's southern magnetic pole.

Just as this bar magnet attracts iron filings to its poles, Earth attracts solar particles above its poles. The north magnetic pole is close to Bathurst Island in northern Canada. The south magnetic pole is in Wilkes Land, Antarctica.

Magnetic Fields

Try holding a magnet on a table and slowly moving a pin (or any steel or iron object) closer to it. At some point, the pin will snap against the magnet. The place where the pulling action begins is the edge of the magnetic field. The magnetic field surrounds the magnet, and its pull is strongest at the magnet's ends, or poles.

Earth's magnetic field extends more than 40,000 miles into space. It reaches so far that in some places the moon passes through it.

17

The Sun also has a powerful magnetic field. As solar-wind particles race through space, they are like miniature magnets, carrying part of the Sun's magnetic field with them. When they reach Earth's magnetic field, the fields push against each other like the opposite ends of two magnets. The particles sweep around Earth's field like water flowing around a rock in a stream. When they arrive at the far side of Earth's field, they drag the field into a long, cometlike tail.

As solar wind passes around Earth's magnetic field, however, many particles become trapped in the field and spiral down toward the planet's magnetic poles. The particles speed along just as electrons in a television picture tube race toward the back of the screen. Earth's upper atmosphere then becomes a screen for auroras.

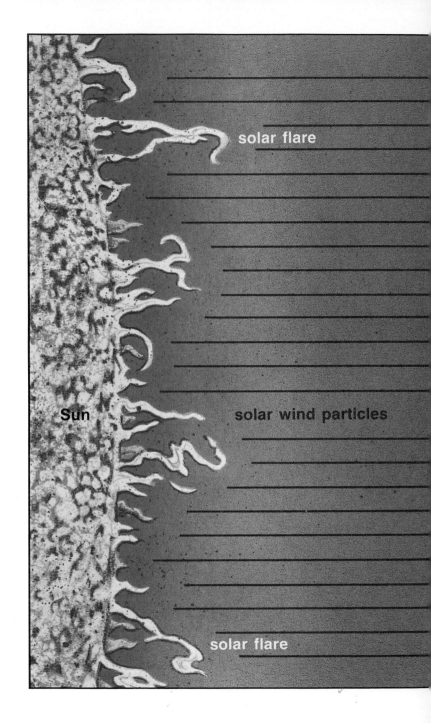

solar flare

Sun

solar wind particles

solar flare

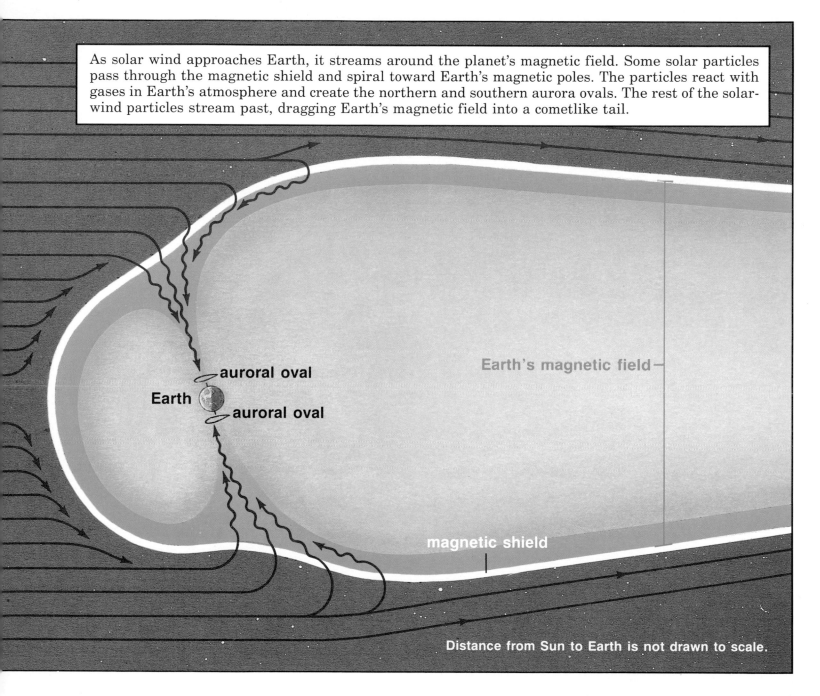

As solar wind approaches Earth, it streams around the planet's magnetic field. Some solar particles pass through the magnetic shield and spiral toward Earth's magnetic poles. The particles react with gases in Earth's atmosphere and create the northern and southern aurora ovals. The rest of the solar-wind particles stream past, dragging Earth's magnetic field into a cometlike tail.

auroral oval

Earth

auroral oval

Earth's magnetic field

magnetic shield

Distance from Sun to Earth is not drawn to scale.

Oxygen and nitrogen gases in the atmosphere above the southern hemisphere are being excited by solar wind. The result is an aurora.

Gases in the Atmosphere

Some of the gases in the atmosphere are made up of free-floating atoms. Others are made up of atoms bound together into molecules, or combinations of atoms. For example, if two atoms of oxygen join with one atom of carbon, they form a molecule of carbon dioxide.

Near Earth's surface, the atoms and molecules of the atmosphere's gases are packed close together, making the gases heavy. These gases make up the air we breathe. Farther above Earth, however, the atoms and molecules spread out, and the gases become so light and thin that we could not breathe.

As solar-wind particles enter Earth's atmosphere, they strike the gases found there. When the atoms and molecules in the gases are hit, some of their electrons escape.

If you have ever taken clothes out of a dryer, you may have noticed how a few pieces cling together. While the clothes twirl around in the dryer and rub against each other, electrons from atoms in a T-shirt, for example, may be knocked from their centers. Atoms in a sock may catch the extra electrons. This makes the T-shirt and sock stick together. When you pull them apart, the electrons crackle and spark as they snap back to where they were.

In a similar way, when the solar-wind protons and electrons smash into the gases in the atmosphere, the atoms become "excited." They move faster, and some of the gases' electrons skip away from their centers. Then they snap back to the way they were, just as those electrons in the T-shirt and sock do when you pull them apart. When they snap back, light is given off. The light given off by the atoms and molecules in the atmosphere is the light of an aurora.

What Do Auroras Look Like?

To the people living near the poles, the lights sometimes look like a faint glow in the sky. Other times, an aurora appears to be a sparkling curtain, a twisting ribbon, a swirling arch, or patches of color. The shape, color, and brilliance of an aurora depend in part on where you are, what time it is, and what is happening on the Sun.

22

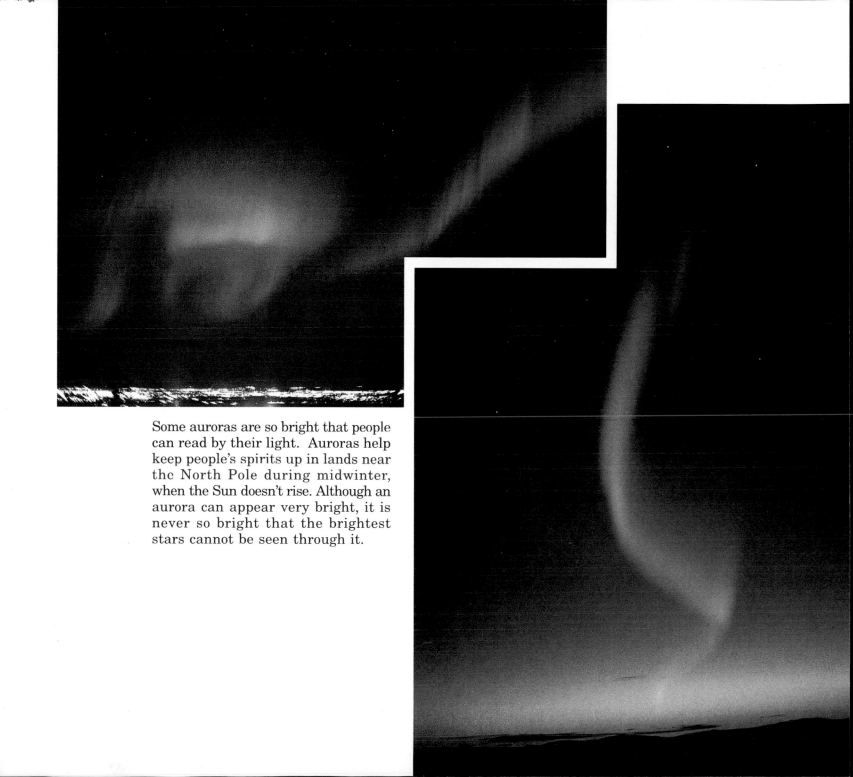

Some auroras are so bright that people can read by their light. Auroras help keep people's spirits up in lands near the North Pole during midwinter, when the Sun doesn't rise. Although an aurora can appear very bright, it is never so bright that the brightest stars cannot be seen through it.

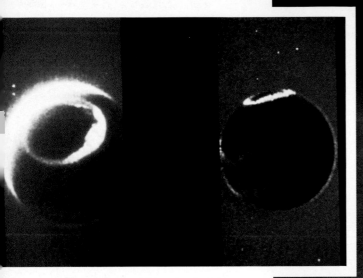

Two views of Earth and its auroras, as seen from a satellite with a special camera *(above)*

Shapes and Movements

If you could soar above Earth in a space shuttle, you would see the lights of the northern and southern auroras in the shape of two huge, lopsided ovals. The upper edges of these ovals are about 1,000 to 2,000 miles above Earth's magnetic poles.

The auroral ovals are present day and night because solar wind is constantly streaming into the upper atmosphere.

As solar-wind particles shoot through the atmosphere, Earth's magnetic field guides them into beams. When the beams hit the atmosphere's gases, columns of light called "rays" are created. These rays sway from side to side.

If you were directly underneath the beams on Earth, you would see bright rays of light shooting out in wavy motions from a point over your head.

Farther away, the rays might look like a giant shimmering curtain curving around the sky.

If you were 200 miles north or south of the curtain, you would see a huge ribbon of light called an "arc." The arc of the aurora would reach from east to west, and like a rainbow, it would look as if its ends were touching the skyline. Actually, the arc and its ends would still be high in the sky.

The aurora would not have changed, but it would look different because of where you were when you saw it.

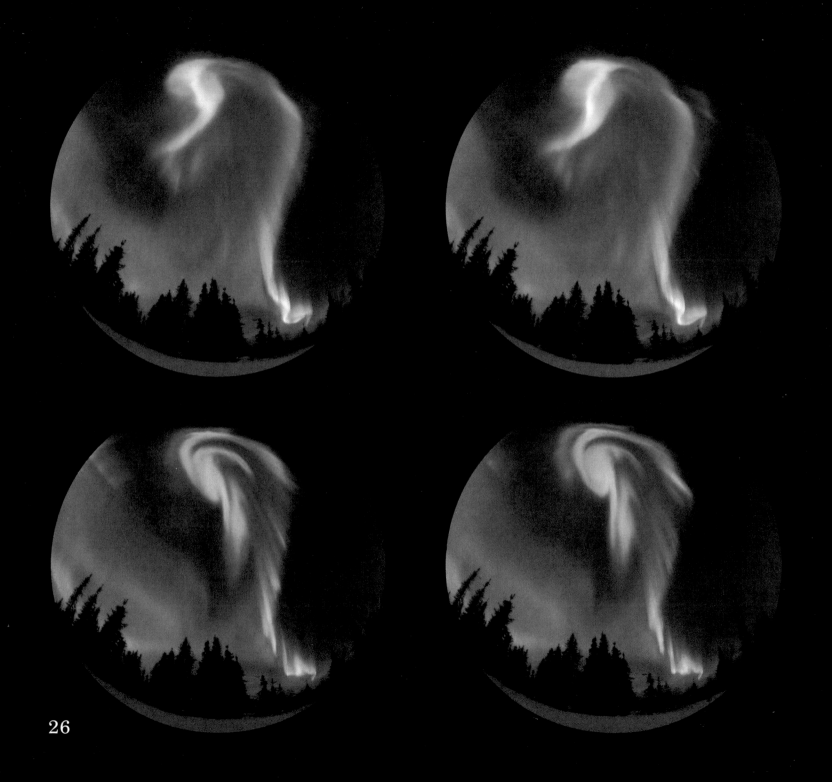

These photographs of the aurora bore- alis were taken a few seconds apart with a special "fish-eye" lens. The movements of an aurora are caused by many forces, including: the rotating of Earth, changes in the amount of solar wind and in Earth's magnetic field, and shifts in the electrical charges in Earth's atmosphere *(opposite)*.

When more particles stream in from the Sun, the rays plunge closer to Earth. They may move in wavy motions or fold and unfold across the sky. Arcs, streaks, and whirlpools may splash the night sky with reds, greens, and sometimes dark purples.

Where do the colors come from?

The answer lies in the atoms and mole- cules of gases in the different levels of the atmosphere.

Colors

Have you ever walked past a sign made with glass tubes of dazzling red? This was probably a neon sign. Neon signs are made by removing all the air from inside a glass tube and filling it with a small amount of neon gas (which is found in our atmosphere). The tube is sealed and wired for electricity. Then, just as in an aurora, a stream of electrons is sent through the tube. When the stream strikes the gas, the electrons in the atoms of the gas become excited. Then they snap back to their normal state. In the process, they give off a glowing red light.

Different kinds of gas shine with different colors when they are struck by electrons. Not all neon signs have neon gas in them. If a mixture of argon and mercury gases are sealed in the tube, a green, blue, or gold light shines. If helium gas is in the tube, it glows with a silver light.

Gases in the atmosphere glow with various colors when they are hit by electrons, just as the gases in neon signs do. The colors that appear depend not only on which gases are struck, but also on how far they are from Earth.

When the electrons in solar winds spiral down near Earth, they may strike oxygen atoms that are about 150 to 100 miles from the ground. Then a deep red lights up the dark night sky. If the solar wind hits nitrogen molecules at this level, a bright pink appears.

Closer to Earth, but still 10 to 16 times higher than jets travel, oxygen atoms flash a greenish white, greenish yellow, or green light when struck. Since oxygen atoms are more "excitable" than other atoms in the atmosphere, this is the most common aurora color. Nitrogen molecules at 80 to 100 miles from Earth tend to glow with blue and violet light. These colors, though, are hard to see against a dark sky.

When electrons dive between 80 and 60 miles from Earth, nitrogen and oxygen molecules color the lower part of the sky with pink bands.

Other gases, such as neon, helium, hydrogen, and argon, also shine with different colors at various distances from Earth. But only small amounts of these gases are found in the atmosphere, and the colors they give off are rarely seen.

When Are Auroras Most Brilliant?

Auroras shine the brightest when the atmosphere is hit with enormous charges of solar wind.

When sunspots appear and solar flares erupt on the Sun, far larger amounts of solar wind strike the atmosphere than normal. At these times, auroras are brighter and can be seen in more places.

Sunspots

Not all parts of the Sun are equally hot. Areas that look dark on photographs are really cooler places. They are called sunspots.

The number of sunspots is always changing, and they change in a pattern. At first, only a few sunspots appear. The number increases over time until there may be more than 100. Then the spots begin to disappear again. The greatest number of sunspots occurs every 11 years. This pattern of increasing and decreasing sunspots is called a solar cycle. We are now in a solar cycle that began with a low number of sunspots in 1986. The number of sunspots increased in 1989 and again in 1991.

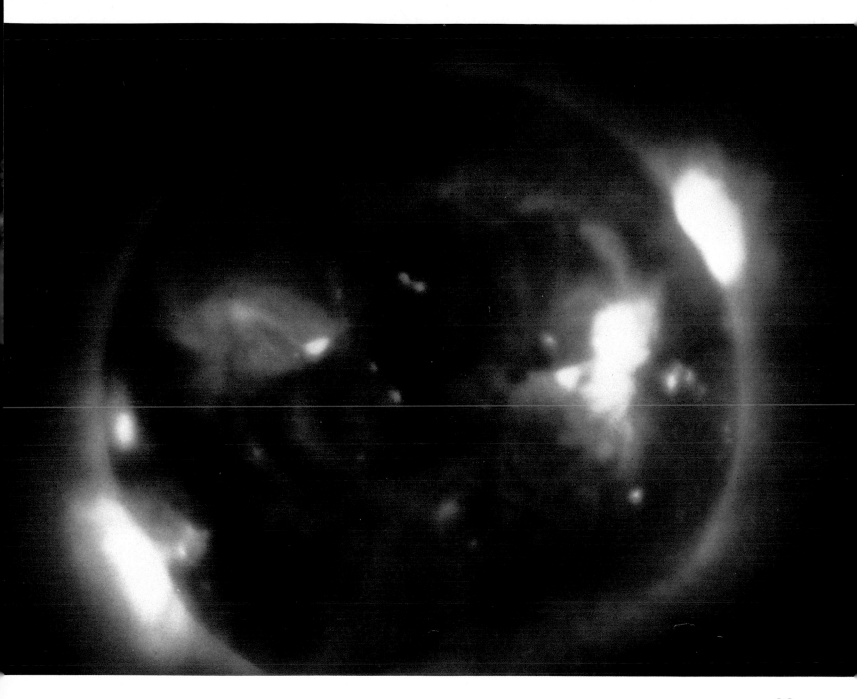

Solar Flares

When sunspots multiply, gigantic solar flares explode nearby. The flares burst from the Sun's surface and flame up like pools of burning gasoline. One flare may be as large as 12 Earths.

Clouds of hot gases billow out, and geysers of solar wind blast through space. When hundreds of millions of tons of these particles plunge into Earth's atmosphere, atoms and molecules of gases are struck again and again. Then even people living far from the magnetic poles are treated to spectacular light shows.

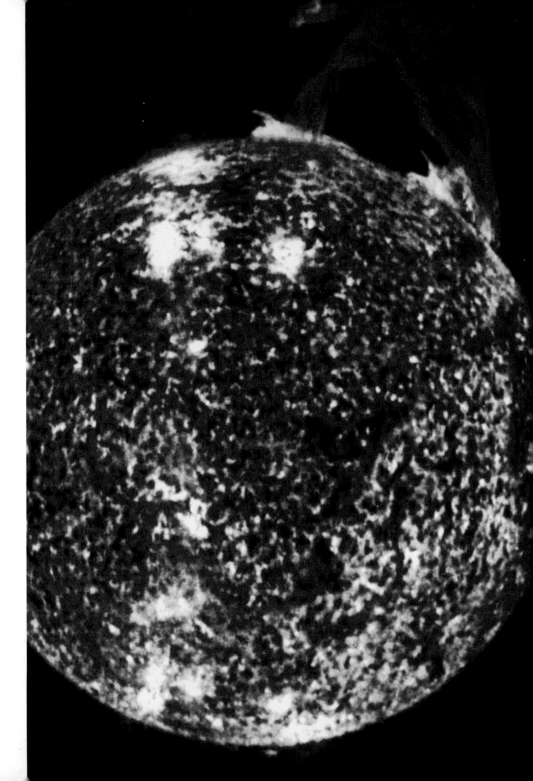

When and Where Can the Northern Lights Be Seen?

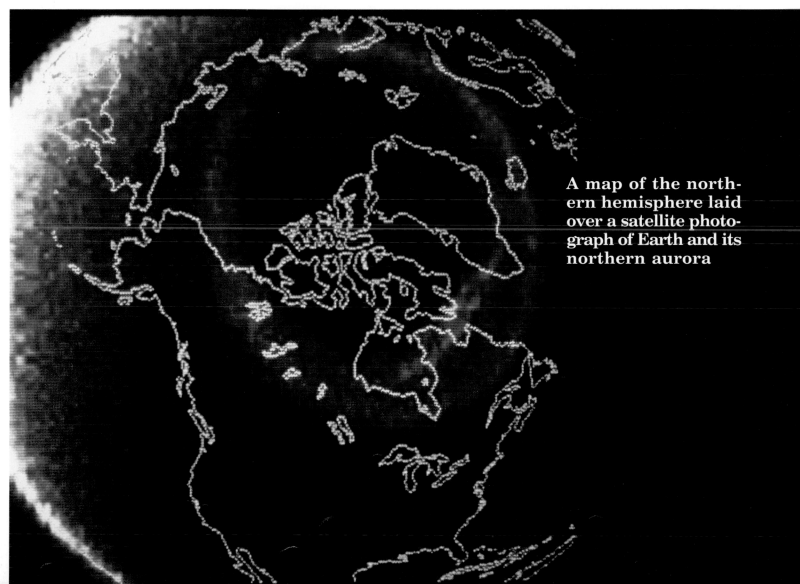

A map of the northern hemisphere laid over a satellite photograph of Earth and its northern aurora

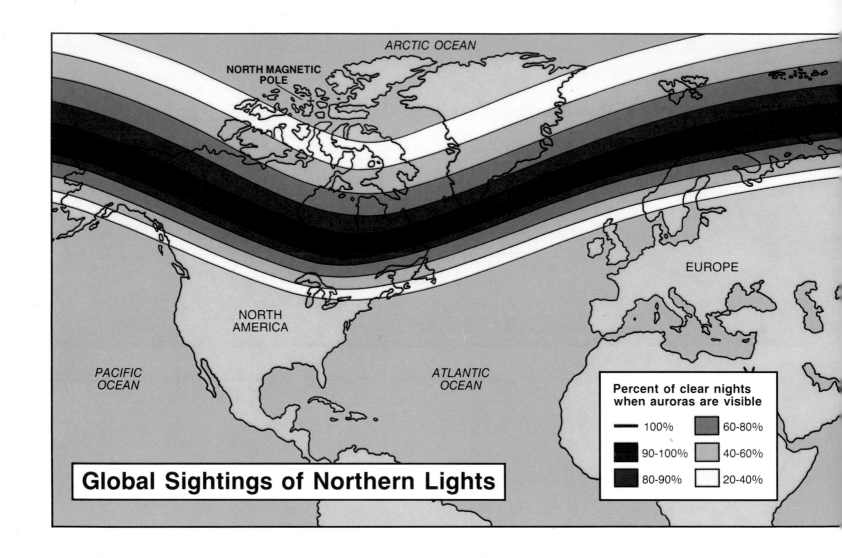

Global Sightings of Northern Lights

Percent of clear nights when auroras are visible

—— 100%		60-80%
90-100%		40-60%
80-90%		20-40%

Not everyone is able to enjoy the northern lights year-round. The farther north you go in North America, the more likely you are to see them. However, even people living in areas where the lights appear frequently do not usually see them during the day. They cannot see them for the same reason they cannot see the stars. The Sun's light outshines the aurora.

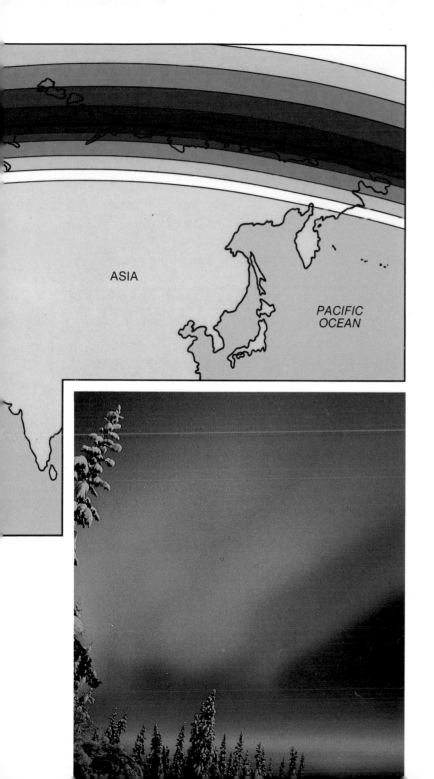

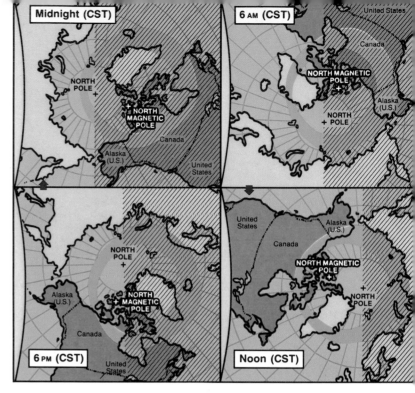

As Earth rotates throughout the day and night, the auroral ovals stay in much the same position in relation to Earth's magnetic poles and the Sun. This diagram shows the position of the northern oval at different times in central North America.

In the evening, people in some northern areas may notice a faint greenish glow near the horizon. Around midnight, the lights brighten as night darkens and North America comes under the wider side of the auroral oval.

37

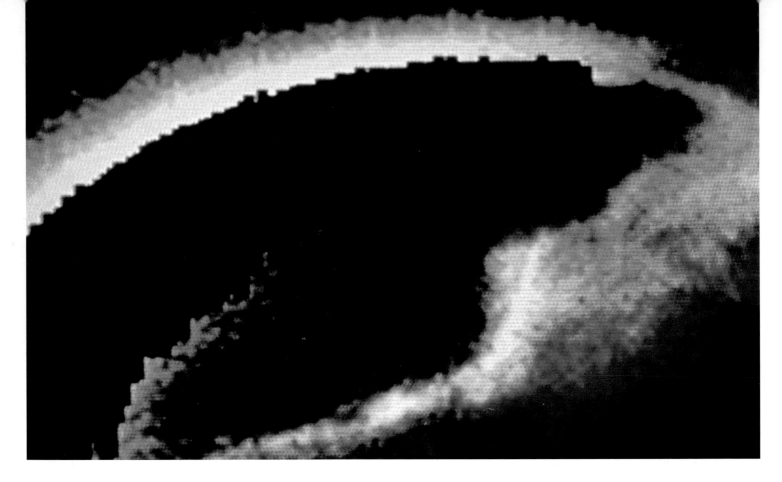

During a Quiet Sun

When the Sun is quiet, with few or no sunspots, the flow of solar-wind particles is steady. The oval above the north magnetic pole is about 2,000 miles wide. People living in Barrow, Alaska, and Churchill, British Columbia, may spot the lights on almost every clear night.

When sunspots appear, more solar-wind particles rush Earthward. The oval widens, and the curtain grows longer. About 18 times a year, the oval becomes so large that it extends over parts of southern Canada, the northern United States, and northern Scotland. People in Seattle and Boston then notice an aurora in the sky.

This computerized photograph shows one of Earth's auroral ovals as seen from space *(opposite)*.

During an Active Sun

At the peak of the solar cycle, when a number of gigantic flares explode from the Sun, sending many more particles than usual toward Earth, the auroral ovals become even wider and larger. The northern oval grows so wide that at times one side of it may extend over the middle of the United States.

The curtain beneath it may be thousands of miles long. It flickers and waves across the sky, and colors change quickly. A reddish glow from the northern lights may even be seen by people living in Mexico.

Sometimes an aurora is bright enough to show up in the early evening and last until dawn. The lights may look blue or purple while the sun is shining. Strong moonlight can also make the lights appear blue.

In November 1991, after a mammoth flare exploded from the Sun, bright lights began filling the sky over parts of the United States as early as 7:00 P.M. A second outburst of colors came around 10:00 P.M. In some places, the aurora lasted until early morning. Even people in Texas, Alabama, Georgia, Oklahoma, and Mississippi witnessed these northern lights. However, sky watchers in Alaska and northern Canada saw nothing. The oval had stretched beyond the point where they could see it.

Danger Signals

While auroras can be breathtaking, they can also signal danger. Storms of erupting solar flares not only make auroras more brilliant, but they sometimes cause unusual things to happen elsewhere. The electrons may block radio and television signals. They can overload power lines, break down telephone systems, and cause huge electric transformers to explode.

In 1989, when a giant flare erupted from the Sun, six million Canadians were left without electricity for nine hours. Power failures shut down parts of Sweden and the United States. Communications were carried to places where they were never meant to be sent. In March, for example, messages of California Highway Patrol officers were heard on radios as far away as Minnesota.

Large amounts of solar wind may also damage one or more of the hundreds of satellites that orbit Earth. Many of these satellites relay weather information or inform scientists about what is happening in space. Some spacecraft keep our armed forces in direct contact with each other. Interruptions in any of these signals or messages can cause dangerous misunderstandings.

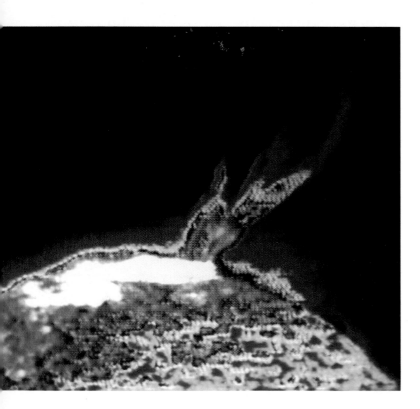

A computerized, heat-sensitive photograph of a solar flare. The heat of the flare increases from the various colored surface areas to the red eruption to the black space, where the heat can be three million degrees. A solar flare sends off unusually high amounts of solar wind.

Studying Auroras

The space age has helped scientists discover many things about auroras and their causes. Satellites have tracked solar wind during periods of both a quiet Sun and an active one. They have measured how much faster and hotter the particles are when flares are erupting on the Sun than when they are not.

In February 1980, scientists launched the satellite *Solar Maximum* to study the Sun. When a flare erupted in May, the satellite recorded temperatures that were hotter than one hundred million degrees.

Other satellites have helped prove the importance of magnetic fields and atmospheres in creating the lights. These satellites sent back photographs that show that planets without magnetic fields, such as Venus and Mars, have no auroras. But Jupiter, which has both an atmosphere and a strong magnetic field, has a large auroral oval above each of its poles. The main gas in Jupiter's atmosphere is hydrogen, and the color it gives off is dark red.

Satellites are being launched during the present solar cycle to find out more about the Sun, Earth, and auroras. The spacecraft *Polar* will orbit Earth at the poles and take pictures of the auroral lights once a minute to see how they change and move. *Geotail* will fly past the moon and examine the tail of Earth's magnetic field. *SOHO* will investigate the Sun, its hot gases, and its particles.

With the help of radar, researchers in such places as Alaska, Canada, Greenland, and Antarctica are gathering information on excited molecules and atoms in the atmosphere. Using magnetometers, they measure the strength of Earth's magnetic field at different times. With photometers, they record just how bright the gases become when they are struck by solar wind.

In addition to studying natural auroras, some experts are creating auroras in laboratories and in space. By examining and measuring what happens during these experiments, they hope to better understand what happens in real auroras.

Some scientists think that in the future a way to harness the power within auroras may be discovered. Then, not only will the gases, solar wind, and magnetic field light up the skies, but they may help to light up our homes.

Tips for Spotting Northern Lights

If you live in southern Canada or the northeastern part of the United States, you may have a chance to see the lights 10 to 75 times a year. But if the Sun is active, you may spot them even if you live farther south.

No matter where you live, it can be discouraging to search for the lights unless you know when to expect them. If you hear a weather report that flares have been exploding on the Sun, follow that tip. Two days later, you can set up watch.

A phone recording at the Space Environment Services Center in Boulder, Colorado, 303-497-3235, can help you decide when to set up a night watch by giving you a K Index. This is a measure of the disturbance in Earth's magnetic field. If the "K" is five or higher, you can expect to see an aurora well south of Canada.

The best place for a lookout is on a hilltop in the countryside away from city lights. Your chances of seeing something are better if the night is clear and moonless.

Let your eyes get used to the darkness before beginning your search. A glow may appear near the horizon, and you may overlook it, thinking it is only the lights of a distant city. *But keep watching.* The most beautiful displays can usually be seen close to midnight. If a giant flare has erupted on the Sun, you may see splashes of color earlier.

If you want to take pictures, a 35-mm camera and high-speed film work best. Bring along a tripod or something to keep your camera steady.

When the lights begin to streak the sky, try different exposure times (between 5 and 60 seconds). Take as many pictures as you can, and you will have a better chance of capturing a lasting reminder of one of Earth's most fascinating displays.

Fascinating Facts

Some people have reported hearing crackling, swishing, snapping sounds during an aurora. Scientists believe these noises may be coming from winds or shifting ice and snow. But until someone makes a recording, no one can be certain.

The Yupik people of St. Lawrence Island say that the lights used to have no colors. Children were warned to stay in at night so they wouldn't be stolen by the lights. Some children didn't listen, and they were carried away. The colors we see now are the colorful parkas of the children as they dance in the skies.

45

One night during the rule of the Roman emperor Tiberius (42 B.C. to A.D. 37), the sky glowed a bright red. The emperor thought the seaport of Ostia on the Tiber River was on fire. He sent his army to put it out, but the men could find nothing but a blazing sky.

Some Inuit peoples in northern Alaska believed the lights were alive and if you whistled at them, they would come closer and snatch you away. Or they might come and cut off your head. Children were warned not to whistle at the lights.

The Menominee people of Wisconsin believed the lights were torches used by great friendly giants of the North to help them spear fish at night.

The Maoris of New Zealand thought the southern lights were reflections of huge fires that had been built by ancestors who settled far south, in cold Antarctica.

46

Natives of the Faeroe Islands off the coast of Iceland warned their children never to leave home without wearing a cap. They feared the lights would burn off their hair.

Children throughout Alaska, such as these in Ruby, Alaska, have been warned about whistling at the northern lights.

During powerful solar storms, animals sometimes become disturbed or act strangely. Homing pigeons, for example, have a hard time finding their way home.

Glossary

Arc: An aurora in the shape of a huge ribbon of light, looping across the horizon

Atmosphere: An ocean of gases, water vapor, and dust particles that extends several hundred miles above some planets, such as Earth

Atom: The smallest particle that can be recognized as being a specific kind of solid, liquid, or gas

Aurora australis: The aurora that appears above the South Pole

Aurora borealis: The aurora that appears above the North Pole

Auroral ovals: Giant, lopsided rings of light above the magnetic poles. The ovals are always present above Earth.

Auroral rays: Beams of light that extend toward Earth from the auroral ovals

Electricity: The movement of electrons from one place to another

Electrons: Invisible particles that normally whirl around the center of an atom

Excited atoms: Atoms with extra energy. Their electrons skip away from the atoms' centers.

K Index: A measurement of disturbances in Earth's magnetic field

Magnetic field: The area around a magnet where the pull of the magnet can be felt

Magnetic pole: The part of a magnetic field where the pull of the magnet is strongest

Magnetometer: An instrument that measures the strength of a magnetic field

Molecule: Two or more atoms joined together

Neutrons: Particles that together with protons make up the center of an atom

Photometer: An instrument that measures the brightness of light

Protons: Particles in the center of an atom

Solar cycle: The 11-year pattern of increasing and decreasing sunspots

Solar flare: A hot explosion of gases on the Sun

Solar wind: Invisible particles—mostly electrons—that regularly stream from the Sun and race through space

Sunspot: An area of the Sun that is cooler and darker in appearance than the rest of the Sun

Index

arc, 25, 27; definition of, 25
atmosphere, of Earth, 12, 18, 21, 24, 32, 34; gases in, 12, 13, 19, 20, 21, 27, 28–31; of Jupiter, 41
atoms, 14, 21, 27; definition of, 15
Aurora, Roman goddess of dawn, 9
aurora australis, definition of, 11
aurora borealis, definition of, 9
auroral ovals of Earth, 19, 24, 37, 38, 39; of Jupiter, 41

beliefs and legends, 8, 45, 46
Boreas, Roman god of wind, 9

colors, causes of, 22, 27–31, 39, 41, 45, 46
Cook, Captain John, 11

electrons, 13, 15, 18, 21, 28, 29, 31, 40
excited atoms, 20, 21, 28, 42

Gassendi, Pierre, 9

K Index, 44

magnetic field, of Earth, 16, 17, 18–19, 24, 41, 42, 44; of Jupiter, 41; of Sun, 18; studying, 44
magnetic poles, 13, 16, 17, 18, 19, 24, 38
magnetometers, 42
molecule, 21, 27, 30, 31, 34; definition of, 21

neon, 28, 31
neutrons, 15
nitrogen, 20, 29, 30, 31
noise, 45

oxygen, 20, 29, 30

photographing auroras, 41, 44
photometers, 42
protons, 15

rays, 24, 25, 27; definition of, 24

shapes of auroras, 22-25, 26-27
solar cycle, 32, 39
solar flares, 18, 32, 34, 39, 40, 41, 44
solar storms, 46
solar wind, 15, 16, 18-19, 20, 21, 24; damage from, 41; definition of, 15; during sunspots, 32, 33, 34, 38; speed of, 16; studying, 41. *See also* solar flares; solar storms
southern lights, 11, 12, 18, 19, 20, 24
Sun, 12, 13, 14, 27, 32, 34, 36, 38, 39, 40; heat of, 14, 15, 41; studying, 14, 27, 32, 41, 44; sunspots, 32, 38